Combating Climate Change. Fisheries and Aquaculture in Zambia

Charles Nyanga

Bibliographic information published by the German National Library:

The German National Library lists this publication in the National Bibliography; detailed bibliographic data are available on the Internet at http://dnb.dnb.de.

ISBN: 9783346208194
This book is also available as an ebook.

Nymphenburger Straße 86
80636 München

Print and binding: Books on Demand GmbH, Norderstedt, Germany
Printed on acid-free paper from responsible sources.

GRIN web shop: https://www.grin.com/document/912136

MINISTRY OF EDUCATION AND TRAINING

NHA TRANG UNIVERSITY

INTERNATIONAL MASTERS IN ECOSYSTEMS MANAGEMENT AND CLIMATE CHANGE

Assignment

Review research papers to identify and analyze problems caused by climate change and variability on fisheries/aquaculture/marine ecosystem in your country. How can you deal with these problems?

Charles Nyanga

Due date: 5th November, 2016

Table of contents

ABSTRACT

The variability in climate has caused a lot of problems to both marine and fresh water aquatic ecosystems. These problems have spread to terrestrial ecosystems posing problems to fresh water aquaculture, forest ecosystems as well as agricultural ecosystems. Climate change has thus become a regional and global problem causing concerns over the preservation of these ecosystems for future generations. One the results of climate change is global warming. It has been predicted that between 2010 and 2070 the world surface temperatures will increase by between 1^0 C to 2.7^0 C. This impact will affect the world economically, socially, environmentally and ecologically. This paper reviews works by experts as well as national and international organizations on impacts of climate change on aquaculture and fisheries in Zambia. It seeks to investigate the methods of mitigating and adapting to climate change which the Zambian government and stake holder organizations have put in place and may be able to put in place.

Further the paper assesses the hindrances to adaptation and mitigation of climate change impacts in Zambia.

Key words: wild capture fisheries, aquaculture, climate change, adaptation, mitigation

INTRODUCTION

Zambia is a landlocked country in Southern Africa, neighboring the Democratic Republic of the Congo to the North, Tanzania to the north-east, Malawi to the east, Mozambique, Zimbabwe, Botswana and Namibia to the south, and Angola to the west. The capital is Lusaka in the south-central part of the country. Currently the total population is estimated to be 15.5 million. Most of the population is concentrated in the Lusaka and the copperbelt province which are the economic hubs of the country.

Zambia with a total land mass area of 752,618 km^2 , has 15 million hectares of water in the form of rivers, lakes and swamps, which provide the basis for extensive freshwater fisheries. Demand for domestic fish for consumption however still outstrips production.

According to Fishakathon (2016), though Zambia is a landlocked country, it has more than 40% of the water resources in the Southern African Development Community (SADC) yet imports an estimated 50,000 tons of fish annually. Water accounts for 20% of the total land mass in Zambia, but yet fisheries and aquaculture accounts for less than 1% of the GDP with more than 300,000 fishers, fish farmers and processors. While it is seen that this is a small percentage, fisheries are crucial to Zambia's rural economy as a source of income and protein. Further Fishakathon (2016) notes that the current national demand for fish is estimated at 120,000 tons per annum but only about 70,000 tons is supplied leaving a deficit of 50,000 tons which is supplied through imports.

In 2015, the Auditor-General's report said that Zambia's fish stocks are multispecies stocks and there is an estimated 400 documented species of which 15 species are commercially exploited. According to DoF, most of the other species are of ecological importance to maintain biodiversity and a clean environment.

The complete picture regarding fish farming in Zambia is still very unclear. The most recent estimates by FAO (2016) give 313 ha as the total hectarage of surface water under production including 47 ha or 15% in government stations, 130 ha or 41% in commercial farms and the rest 44% in rural ponds exploited mostly for domestic/substance purposes. This area estimate of 313 ha represents a substantial increase from the 1967 estimate of 100 ha.

There are 19 government stations, with a total of 338 ponds, of 1,400 m^2/pond on average. The number of commercial fish-farmers is estimated at 90. Total number of ponds is estimated at 500, with an average of 3,600 m^2 / pond (including small reservoirs stocked for fish-farming). There would be around 2000 rural farmers which have included fish-farming in their activities, they would exploit 2,162 ponds with an average of 400 m^2/pond.

FAO (2016) estimates that the average yield for the government stations is 2 tonnes/ha which represents a total of 94 T for that sector. Commercial fish-farmers are estimated to have average yields of 3 tonnes/ha which would give a total production of 540 tonnes. Rural ponds have estimated yields of 1 tonnes/ha, for a total production of 86 tonnes. The total production from fish farming in Zambia is estimated at 710 tonnes. This represents a considerable increase from an estimated total production of 88.7 tonnes in 1967.

The scenario regarding capture fisheries is also not very clear. The Luapula Basin consists of the Chambeshi River, the Bangweulu Lakes and surrounding swamps and the Lake Mweru-Luapula fishery. The Zambezi catchment consists of the Luangwa River, Lukanga Swamps, Kafue River, Upper, Middle and Lower Zambezi. The middle Zambezi is now dominated by Lake Kariba. Other fisheries are Lake Tanganyika and the Mweru-wa-Ntipa fishery. These fisheries produce an estimated 70,000 – 80,000 metric tonnes of fish per annum. On employment contribution, about 20,000 people are directly employed in the industry and 250,000 are involved in fish processing, trading and subsistence fishing. The sector's contribution to Gross Domestic Product (GDP) is estimated at 3.8% and it is the third largest employer after crop production and mining (MACO, 2008)

The history of fisheries production in Zambia is provided by Table 1 below. The fisheries production in most cases has been stagnant except for Bangweulu and Lusiwasi fisheries which have shown a drastic drop.

	Mweru Luapula		Bangweulu		Itezhi tezhi		Lusiwasi		Kariba		Tanganyika		Mweru wantipa	
	1997	2013	2007	2012	2006	2014	2006	2014	2006	2011	2004	2011	2004	2013
Production (Metric tonnes)	8,964	12,187	15,098	13,573	2,007	2,033	1,933	1,200	8.,008	9,454	13,364	15,953	3,064	3,416
Level of effort (by count)														
Boats	8,662	13,977	11,281	12,103	1,129	520	164	240	2,431	2,451		2,320	1,798	3,535
Fishers	12,047	20,936	15,113	18,150	1,172	1,065	365	431	2,004	4,653		8,420	2,337	4,929
Gear	97,303	263,630	56,296	59,663	11,541	8,297	856	2,387	17,102	26,769		16,160	73,367	20,854

Table 1: Showing production and effort applied per Fishery (Source: DoF, Zambia, 2015)

Given the foregoing, the objectives of this paper are:

- To identify the climate change scenarios
- To analyze the problems of climate change
- To identify the possible adaptation and mitigation strategies which Zambia may apply.

The methodology adopted is reviewing relevant literature and reviewing papers and other works on climate change in Zambia and the adaptation and mitigation strategies.

POTENTIAL IMPACTS OF CLIMATE CHANGE ON FISHERIES AND AQUACULTURE IN SUB-SAHARAN AFRICA

Impacts of climate change in Sub-Saharan Africa

The term climate change is used to describe a marked change in the long-term average of a region's weather conditions. Climate is the average weather together with its variability. It is therefore the synthesis of weather in a given place over a period of at least 30 years (Akinsanmi, 2009). Climate parameters include temperature, rainfall, humidity, dew, wind, sunshine, mist, haze and clouds. Thus a drastic and permanent departure of climate patterns from mean values observed, constitute climate change.

Mohammed and Uraguch (2013) wrote that Africa is considered to be a land of plenty with massive inland and marine water resources. The continent has been deemed " the continent most vulnerable to impacts of projected climate change" by the UN Intergovernmental Panel on Climate Change. The continent has one of the most volatile water systems on the planet, and its rivers routinely experience wild swings in flow. For example, variation in the Zambezi River is estimated to be ten times higher than that of most European rivers. This situation has been worsened with climate change. Additionally, climate change is expected to increase the extremes of drought and flooding, with the result that Africa's already highly variable climate and hydrology will eventually be even more difficult to predict, making fisheries and aquaculture activities very difficult to plan and manage.

Mohammed and Uraguch (2013) , additionally have noted that the major aquatic habitats in Sub-Saharan Africa (SSA) include the Lake Malawi, Lake Victoria, Lake Kariba, river Nile, river Zambezi and many coastal estuarine, deltas, floodplains, mangrove swamps and inland wetlands.

The diversity of the habitats and the species they support respond differently to different impacts of climate change. The climate change is very likely to lead to fluctuations in fish stocks and these will lead to major economic consequences for many vulnerable communities and nations that depend on fisheries.

On the impacts on fish stocks , Mohammed and Uraguch report that the impacts of climate change on fish stocks in SSA can be classified as physical and biological changes. Physical changes include surface temperature rise, sea level rise, changes in salinity and ocean acidification. Biological changes include changes in primary production and changes in fish stock distribution. These impacts when combined together will have adverse impacts on the already strained resources.

The IPCC (2007) , which has been very instrumental in advocating for climate change issues by bringing together governments, ENGO, stakeholders and financiers had summarized the projected overall impacts in Africa due to climate change in Panel 1 shown below.

Summary of the Projected Impacts of climate change in Africa

- By 2020, between 75 and 250 million people in Africa are projected to be exposed to increased water stress due to climate change.
- By 2020, in some countries, yields from rain-fed agriculture could be reduced by up-to 50% Agricultural production, including access to food, in many African countries is projected to be severely compromised. This would further adversely affect food security and exacerbate malnutrition.
- Towards the end of the century, projected sea level rise will affect low-lying coastal areas with large populations.
- By 2080, an increase of 5 to 8% of arid and semi-arid land in Africa is projected under a range of climate scenarios (TS).
- The cost of adaptation could amount to at least 5 to 10% of Gross Domestic Product (GDP).

Sources: Report Summary for Policy Makers, IPCC, 2007

Panel 1: Summary of potential impacts of climate change in Africa

POTENTIAL IMPACTS OF CLIMATE CHANGE ON FISHERIES AND AQUACULTURE IN ZAMBIA

Climate change impacts can be classified as:

- Catastrophic – these are climate disasters or hazards such as typhoons, hailstorms, droughts and sudden floods.
- Chronic – these are new conditions such as higher temperatures, sea level rise, saline intrusions, subsiding water tables, more or less rainfall, less predictable seasons.

Climate change impacts on the environment and aquatic ecosystems in Zambia

Zambia is endowed with considerable environmental assets, including 50 million hectares of forest and a rich wildlife estate and protected area system covering some 36% of the total land mass. These natural resources are major contributors to the GDP (e.g mining, tourism, agriculture, forestry) and amounts to 27% of national wealth in comparison to 2% in Overseas Economic Cooperation and Development countries in Africa receive and are critical to Zambia's effort to achieve sustainable development.

This scenario, however, is being affected by climate change. Climate change had already lead to changes in terrestrial, freshwater and marine ecosystems in Zambia. The extreme weather patterns have demonstrated the vulnerability of some of Southern Africa's ecosystems, Zambia inclusive. The migration patterns, geographic range and seasonal activity of many terrestrial and marine species have shifted in response to climate change. The abundance and interaction among species has also changed (IPCC, 2014). Furthermore, there appears to be increased growth of certain species more than others, the increased growth has produced knock-on ecological impacts; for example there seems to be an entire shift of certain biomes with certain grasslands being taken over by woody ecosystems altogether. Despite the fact that the African continent has contributed the least to anthropogenic factors of climate change, Africa is the worst hit and will pay the highest price for climate change.

Climate change impacts on fisheries in Zambia

The productivity of a fishery is tied to the health and functioning of the ecosystems on which it depends for food, habitat and even seed dispersal (MAB, 2009).Generally the only control humans can exert over a fishery's productivity is adjustment of fishing effort (Brander, 2007). Estuaries, mangroves, coral reefs and seagrass beds are particularly significant in the provision of ecosystems services, especially as nurseries for baby fish, and are amongst the most sensitive and highly exposed to the negative impacts of coastal development, pollution, sedimentation, destructive fishing practices and climate change. Fish tend to live near their tolerance limits of a range of factors; as a result, increased temperature and acidity, lower dissolved oxygen and changes to salinity can have deleterious effects (Roessig et al.,2004). In particular the changes in climate which will have an impact on the fisheries in Zambia will be characterized by alterations in average temperature and primary production.

Increase in average temperature

All marine and aquatic invertebrates (molluscs, crustaceans, worms etc.) and fish are poikilotherms; their internal temperature varies directly with that of their environment. This makes them very sensitive to changes in the temperature of their surrounding environment. When changes do occur they move to areas where the external temperature allows them to regain their preferred internal temperatures. Therefore the tendency which has been predicted is that the cooler regions will gain while the tropics , like the Zambezi, will suffer losses of upto 40% (Cheung et al., 2010). This is the species migration effect of increases in temperature.

Another impact of changes in average temperature is reported by Walther et al., (2002). This involves recruitment which is reduced due to higher temperatures in the aquatic environments. Eaterling et al., (2007) also reported that some stocks may fall below sustainable levels due to alterations in temperature even if the stocks are subjected to the same fishing effort. Where fish continue to inhabit waring bodies of water the increase in temperature will increase their metabolic rate hence slowing growth and consequently reducing maximum species length and weight.

Primary Production

Primary production is affected by availability of nutrients in the water, which in turn depends on freshwater run-off and ocean mixing as well as levels of light and temperature. In some areas reduced precipitation could lead to reduced run-off from land, starving wetlands and mangroves of nutrients and damaging local fisheries. In other areas increased precipitation or increased extreme weather events, including flooding, will lead to excessive nutrient levels in rivers, lakes as sewage and fertilizer is washed into water bodies causing harmful algal blooms (Roessig et al., 2004). Primary productivity is predicted to decline at lower latitudes where most of the world's small scale fisheries are located, reducing the productivity of the fisheries. (FAO. 2008a)

Other effects of climate change on fisheries in Zambia

Other impacts of climate change, which are a combination of chronic and catastrophic effects, on the fisheries in Zambia are:

- Increased frequency of extreme weather events which affect safety of fishers, damage homes, services and infrastructure and aquatic ecosystems
- Increases in unpredictably heavy rainfall events will increase flood risk, reduced water quality, threaten infrastructure
- Reduced dry season flows in African rivers generally and Zambian rivers in particular will result in reduced fish yields due to impacts on spawning and larval dispersion (UNEP-WCMC, 2006)

Climate change impacts on aquaculture in Zambia

The climate change impacts on aquaculture are much more complex to study as compared to climate impacts on wild fisheries. The main impacts can be seen on the production environment. In aquacultural activities in most tropical countries, Zambia inclusive, greater control can be exerted over the production environment, e.g by providing food, controlling breeding and diseases, and over environmental conditions, e.g controlling water flows, temperature, water quality and many other parameter. This has brought the tendency of depending less on ecosystem services. However many small-scale fish farmers in Zambia practice low-input, low-output form of aquaculture depending heavily on ecosystems services and naturally available feed to support their fish. Many fish farmers depend on wild fish stocks for food and seed, supply of fishmeal and oils from capture fisheries, used as feed stock in aquaculture is declining due to climate change impacts on wild fisheries (IPCC, 2007a; Naylor et al.,2000). Also increased run-off bringing in nutrients from sewage or agricultural fertilizers as well urban waste may cause algal blooms which in turn lead to reduced levels of dissolved oxygen and fish deaths (Diersing, 2009).The rise in temperatures similarly reduces levels of dissolved oxygen and increased metabolic rates of fish, leading to increases in fish deaths, declines in production or increases in feed requirements while increasing the risk and spread of disease (FAO, 2008a).

CLIMATE CHANGE ADAPTATION AND MITIGATION STRATEGIES

Adaptation is key in context , defined as adjustment to natural or human system in response to experience or future variability and extreme events which may be beneficial or adverse (NAPA, 2008).

Mitigation, therefore, means taking actions before, during and after a disaster has happened in order to reduce the negative effects of disasters (NAPA, 2008).

Policy framework and civil society participation in Climate Change awareness

Banda (2011), writing for JCTR, observed that in terms of policy response to climate change Zambia has a fair number of both public institutions and NGOs participating in national forums and awareness programmes on climate change. The participation has ranged from discussing mitigation strategies to adaptation. Banda (2011) further goes on to list the main players here as:

- Climate Change Facilitation Unit (CCFU) under the Ministry of Tourism and Environment – this organisation was established in 2009 by the Ministry of Tourism, Environment Natural Resources with support from UNDP and Norwegian government . It is an extended arm of government to spearhead on-going climate change activities.
- Zambia Civil Society Climate Change Network (ZCSCCN) – was established in 2009. It encompasses seven core organisations and a network of over fifty individuals and organisations working in the field of climate change and natural resources management. Its objective is to promote the sharing of climate change and climate related information among all stake holders to enhance advocating and engagement in climate change at national, regional and international levels.
- Zambia Community based Natural Resources Management(ZCBNRM) Forum – the ZCBNRM Forum was formed in 2004 with support from World Wide Fund (WWF) Zambia office. The ZCBNRM forum sits on the REDD+ (Reduction of Carbon Emissions and Forest Degradation and Destruction) steering committee as a representative of civil society.
- National Climate Change and Development Council (NCCDC) – also has activities aligned to climate change
- National Climate Change Reduction Strategy (NCCRS) – also has activities aligned to climate change

Recommended climate change mitigation and adaptation strategies for Zambia and Sub-saharan Africa

In consideration of climate change in fisheries and aquaculture in Africa Professor Eyimunmi Falaye (2013) in his write up for the African Union Inter-Africa Bureau on Animal Resources has made suggestions for the following adaptation and mitigation strategies to African governments generally and the Zambian government may adopt the following climate change adaptation and mitigation strategies as well:

- There should be national and local depositories of climate change and related data which may be easily accessed by researchers, development practitioners, fish farmers and other stakeholders,
- The government should ensure increased capacity in the areas of information technology and modelling in all sectors of the economy to better address vulnerability and build resilience at the local, provincial and national levels
- Suitable adaptation and mitigation measures should be site specific to respond to anticipated changes in rainfall and temperature in Zambia.
- Efficient natural water way management systems should be advocated for,
- Appropriate protection strategies should be put in place for watershed and wetland conservation,
- Mainstream gender in all relevant policies, with greater involvement of women in decision making in climate change issues
- Collective efforts by the government, NGOs, CBOs and the private sector to create awareness and education on environmental management matters across all segments of the nation
- The government should facilitate capacity building in sustainable aquaculture and fisheries techniques with respect to climate change adaptation and mitigation
- Models for sustainable fisheries and aquaculture management and aquatic resources conservation would be required to be in place for regeneration of fishery stocks and ecosystems.
- There should be serious scientific research on integrated management of water resources.

- The government should have in place sound policy measures specifically designed to cover resilience, mitigation, adaptation and management of climate change impacts on fisheries and aquaculture.
- Information on scientific findings relating to climate change should be translated into local languages to facilitate common understanding.

CONCLUSION

In conclusion, it may be said that the key objectives which were : (1) to identify climate change scenarios; (2) analyze the problems of climate change in fisheries and aquaculture; (3) identify the possible adaptation and mitigation strategies have been met by analyzing some papers and publications and conference proceedings reviewed, all the works point to the same findings about the impacts of climate change on ecosystems, wild fisheries and aquaculture. These impacts are quite severe and costs will rise to the African governments in general and Zambian government in particular. The publications also all emphasize that governments should take action now if they are to save the aquatic ecosystems and terrestrial ecosystems.

ACKNOWLEDGEMENTS

The student (Charles Nyanga), wishes to extend sincere gratitude to Dr Nguyen Lam Ahn, Dr Nguyen Van Minh, Dr. Pham Quoc Hung, Dr. Le Ming Hoang for their patience, understanding and for the wonderful lectures , materials and field trip , to the fish hatchery factory, from which we have acquired a lot of knowledge in Aquaculture and Fisheries. Also thanks go to the classmates for the team work and understanding, and the fellow group members as well for exhibiting team work.

REFERENCES

AMCEN Secretariat, Fact Sheet (2007): Climate Change In Africa - What Is At Stake? *Excerpts from IPCC reports, the Convention,* & BAP
Compiled by AMCEN Secretariat.

Auditor General, (2015): Government of the Republic of Zambia, *Report of the Auditor General on Sustainable Management of Fish Resources in Natural Waters.*

Brander, K.M. (2007): Global Fish Production and Climate Change, *Proceedings of the National Academy of Sciences* 104(50), 19704 – 19714.

Brander, K.M. (2010): Impacts of Climate Change on Fisheries. *Journal of Marine Systems*, 79 (3 – 4), 389 – 402.

Diersing, N. (2009): Phytoplankton Blooms: The Basis. Florida Keys National Marine Sanctuary, Key West, Florida, USA, 2 pp. Available at http://floridakeys .noaa.gov/pdfs/wqpb.pdf.

Easterling, W,E., Aggarwal. P.K., Batima, P. et al. (2007): Food, Fibre and Forest Products.In: *Climate Change 2007: Impacts Adaptation and Vulnerability. Contribution of Working Group II to the Fourth Assessment Report of the Intergovernmental Panel; on Climate Change.* Cambridge University Press, Cambridge, UK, pp. 273 – 313. Available at http:// www.ipcc.ch/pdf/assessment-report/ar4/wg2/ar-wg2-chapter5.pdf.

Falaye, E. (2013): Strengthening the Capacity of Fisheries Communities for Adaptation to Climate Change: *Identification of Policy Intervention Entry Points.* African Union – Inter-African Bureau for Animal Resources.

FAO (2008A): Climate Change Implications for Fisheries and Aquaculture. In: *The State of Fisheries and Aquaculture 2008.* FAO, Rome, Italy, pp. 87 – 91.

IPCC, 2007: *Climate Change 2007: The Physical Sciences Basis. Contribution of Working Group I to the Fourth Assessment Report of the Intergovernmental Panel on Climate Change*, Cambridge University Press, Cambridge, United Kingdom and New York, NY, US, 996 pp.

IPCC (2007a) *Climate Change 2007: Synthesis Report – Contribution of Working Groups I, II, and III to the Fourth Intergovernmental Panel on Climate Change.* Core Writing Team: R.K Puchauri and A. Reisinger, eds. IPCC, Geneva, Switzerland, 8 pp.

Kalantany,.C. (2010): Climate Change in Zambia: Impacts and Adaptability, *Global Majority Journal.*

MAB (Multi – Agency Brief) (2009): *Fisheries and Aquaculture in a Changing Climate*. FAO, Rome, Italy, 6 pp. Available at ftp://ftp:fao.org/FI/brochure/climate_change/policy_broief.pdf.

Mohammed and Uraguchi (2013), Impacts of Climate Change on Fisheries: Implications for Food Security in Sub-Saharan Africa, *Global Food Security Journal,* Nova Science Publishers

National Adaptation and Programme of Action (2008), *Climate Change - Policy Brief,* Government of the Republic of Zambia

Naylor, R.L., Goldburg, R.J., Primavera, J.H. et al. (2000). Effect of Aquaculture on World Fish Supplies. *Nature 405*, 1017 - 1024

Roessig, J.M., Woodley, C.M., Cech, J.J. and Hansen, L.J. (2004): Effects of Global Climate Change on Marine and Esturine Fishes and Fisheries. *Reviews in Fish Biology and Fisheries 14*, 251 – 275.

UNEP – WCMC (2006): *In the Front Line: Shoreline Protection and Other Ecosystem Services from Mangroves and Coral Reefs*. UNEP – WCMC, Cambridge, UK, 33 pp. Available at: http://data.iucn.org/dbtw-wpd/edocs/2006-025.pdf.

Walther, G., Post, E., Convey, P. et al. (2002): Ecological Response to Recent Climate Change. *Nature 416*, 386 – 395.

Williams, L. and Rota, A. (2013): *Impact of Climate Change on Fisheries and Aquaculture in the Developing World and Opportunities for Adaptation*, Fisheries Thematic Paper: Tool for Project Design.

World Bank (2005): *Turning the Tide. Saving Fish and Fisheries: Building Sustainable and Equitable Fisheries Governance*. The World Bank, Washington, DC, USA, 20 pp. Available at: www.seaweb.org/resource/documents/reports_turningtide.pdf.